Weird World of Animals

Animals with Super Stomachs

By Mary Griffin

Gareth Stevens Publishing

Please visit our website, www.garethstevens.com. For a free color catalog of all our high-quality books, call toll free 1-800-542-2595 or fax 1-877-542-2596.

Names: Griffin, Mary, 1978-.
Title: Animals with super stomachs / Mary Griffin.
Description: Buffalo, NY : Gareth Stevens Publishing, 2026. | Series: Weird world of animals | Includes glossary and index.
Identifiers: ISBN 9781482471779 (pbk.) | ISBN 9781482471786 (library bound) | ISBN 9781482471793 (ebook)
Subjects: LCSH: Animal–Juvenile literature. | Anatomy–Juvenile literature. | Digestive organs–Juvenile literature. | Adaptation (Biology)–Juvenile literature.
Classification: LCC QL806.5 G75 2026 | DDC 571.31–dc23
-
Published in 2026 by
Gareth Stevens Publishing
2544 Clinton Street
Buffalo, NY 14224

Cover Designer: Nicholas Switalski
Interior Designer: Rachel Rising
Editor: Therese Shea

Photo credits: Cover, p. 1 Korneliia Lutso/Shutterstock.com; Cover, p. 1, 3-24 LarysaPol/Shutterstock.com; p. 5 Wirestock Creators/Shutterstock.com; p. 7 Clara Bastian/Shutterstock.com; p. 7 Designua/Shutterstock.com; p. 9 AbuElia/Shutterstock.com; p. 11 Natalia Kuzmina/Shutterstock.com; p. 13 yamasan0708/Shutterstock.com; p. 13 (shark) frantisekhojdysz/Shutterstock.com; p. 15 Pallav Pranjal Sarma/Shutterstock.com; p. 16 faiz0023/Shutterstock.com; p. 17 Dennis Jacobsen/Shutterstock.com; p. 19 John Ceulemans/Shutterstock.com; p. 19 Sean Clee/Shutterstock.com; p. 21 Jeffry Surianto/Shutterstock.com.

Printed in the United States of America

CPSIA compliance information: Batch #CSGS26: For further information contact Gareth Stevens, at 1-800-542-2595.

CONTENTS

*Words in the glossary appear in **bold** type the first time they are used in the text.*

SUPER ADAPTATION

Your stomach is so good at its job that you probably don't think about it much. It's the part of your body that stores and breaks down food. It helps you get the **nutrients** your body needs.

Many kinds of animals have stomachs that work like yours. Others have super stomachs! Their stomach may help them eat things that most animals can't. Their stomach may save them from poison. Their stomach might even save an animal from a predator. Read on to learn about animal stomachs that help animals **survive**!

Not all animals have stomachs! The **mammal** called the platypus, shown below, doesn't. Special **adaptations** make this possible.

That's WEIRD!

Many kinds of fish don't have stomachs. They have different ways of digesting food, which means breaking it down into usable pieces.

THE COW'S CUD

Cows have a **complex** stomach. Some people say they have two or even four stomachs! In fact, a cow's stomach has four chambers, or spaces. Here's how it works: A cow chews grasses and other plants. It swallows its food, which goes to two chambers. Then, the cow throws up the food. It chews this "cud" again. When it swallows again, the cud goes to two different chambers.

All four chambers help break down the plants and get the cow as many nutrients as possible from them.

Follow the arrows to figure out how a cow's stomach works!

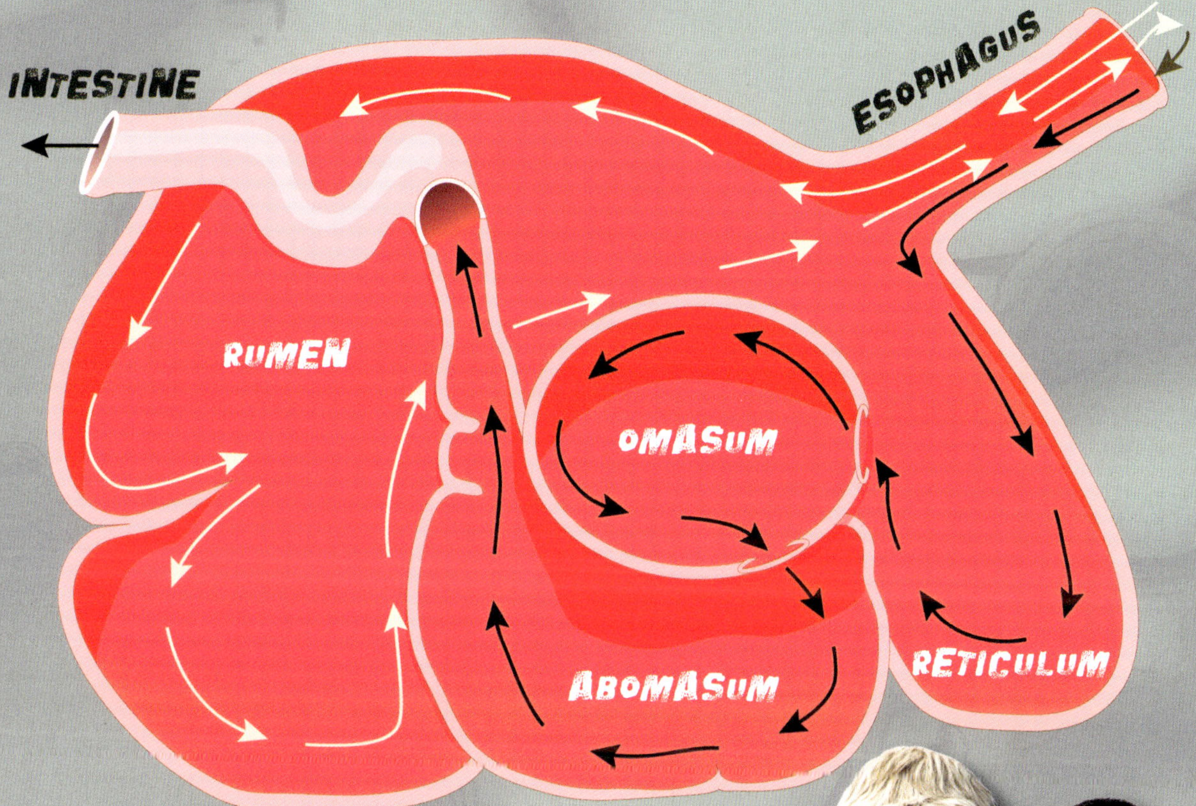

That's WEIRD!

A cow is a ruminant. This is a mammal that chews cud that has been thrown up from its rumen, shown in the picture. Goats, sheep, giraffes, and deer are also ruminants.

CLEVER CAMEL

A camel chews cud too. However, it has a three-chambered stomach instead of four. In its desert home, camels eat whatever plant life they can find. This includes thorns and other tough plant parts.

Like cows, camels regurgitate, or throw up, their food, chew it again, and send it back to their stomach for more digesting. Camels may go days without eating in the desert, so getting as many nutrients as they can from what they do eat is important.

Camels store fat in their humps for times when they cannot find food.

That's WEIRD!

A camel can go a week or more without drinking water!

THE HUNGRY BULLFROG

The American bullfrog eats almost anything that moves! For smaller prey, it uses its sticky tongue to capture the animal and pull it in. The bullfrog may use its front legs to stuff larger prey into its mouth.

A bullfrog doesn't have teeth to chew. The inner lining of the stomach has many folds, so the stomach gets larger when needed. Strong liquids called acids help break down the food in the stomach. Then, the broken-down food moves into another part of the digestive system called the **intestines**.

American bullfrogs are the largest North American frogs. They can weigh 1 pound (0.5 kg). They eat birds, small mammals, and other bullfrogs.

That's WEIRD!

Bullfrogs are named for their call, which some think sounds like a bull.

TOADS' TRICK FOR TOXINS

Many kinds of toads have a clever way of getting rid of toxins, or poisons, in their body. They can't throw up harmful things they've eaten, but they can throw up their whole stomach!

When the stomach comes out, it's turned inside out, spilling out whatever was in there. The toad might even wipe the inside of the stomach with its front feet! Then, it stuffs the stomach back into its body. What a

The cane toad, shown here, is one species that can throw up its stomach. This is called a stomach eversion.

That's WEIRD!

Some shark species can also turn out their stomachs to avoid being poisoned by something they ate.

TIGER SHARK

VULTURES' STOMACH OF STEEL

Vultures are known for eating some really gross things. How does a dinner of garbage and **carrion** sound? Vultures aren't often interested in hunting for their meals. They like to eat animals that are already dead.

Vultures don't just eat dead meat. They eat bones too! Some studies report bones make up about 90 percent of a vulture's diet. That means vultures have to have a really tough stomach to digest their food.

Vultures have excellent eyesight. They've been known to spot carrion on the ground from 4 miles (6.4 km) away!

That's WEIRD!

Vultures may store food in a pouch called a crop before digesting it. If a predator comes near, they may dump this food.

Vultures have the strongest stomach acid of any animal! It's 100 times stronger than the acid in a person's stomach. This acid can break down hard bits of carrion so that nutrients can be absorbed, or taken in, by the bird's body. The acid also easily kills some harmful bacteria, **parasites**, and other deadly things that grow on the rotting things that vultures eat.

Vultures are important to their habitats, or natural homes. They remove dead animals that can spread disease, or illness, to living animals.

Vultures use some kinds of bacteria found on carrion to help them digest too.

HYENAS' NASTY MEALS

Hyenas are one of the few animals that could compete with vultures for the toughest stomach award. And their stomach isn't the only tough body part. Spotted hyenas' tough teeth and powerful jaws can bite through bone!

Hyenas' digestive systems can break down a whole animal's body, including skin, hooves, horns, hair, and bones. Anything that isn't broken down is regurgitated as **pellets**. Hyena waste droppings are whitish because of the great amount of bone matter they eat.

Hyenas sometimes work together to kill larger animals.

That's WEIRD!

Hyenas eat other animals' droppings too!

HYENA DROPPINGS

SO MANY MORE

People have long been interested in ostrich stomachs. Ostriches swallow rocks—on purpose! They help the bird break down food in their stomach. Pufferfish don't use their stomach for digestion. Instead, they use it to puff up their bodies to defend themselves.

Perhaps the weirdest of all is the sea cucumber. It can throw its whole digestive system at predators. After it gets away, it can regrow its insides. Animals like these and many others have **evolved** some really super stomach adaptations!

Pufferfish look like strange, spiky balloons when they puff up!

Glossary

adaptation: A change in a type of animal that makes it better able to live in its surroundings.

carrion: A dead, rotting animal.

complex: Having to do with something with many parts that work together.

evolve: To grow and change over time.

intestine: A tube in the body that helps digest food.

mammal: A warm-blooded animal that has a backbone and hair, breathes air, and feeds milk to its young.

nutrient: Something a living thing needs to grow and stay alive.

parasite: A living thing that lives in, on, or with another living thing and often harms it.

pellet: A small, round mass of something.

survive: To stay alive.

FOR MORE INFORMATION

Books

Chanez, Katie. *Gross Animal Habits*. Minneapolis, MN: Jump!, Inc., 2025.

Taylor, Barbara. *Inside Animals*. Beverly, MA: Wide Eyed Editions, 2021.

Websites

Spotted Hyena Facts!

www.natgeokids.com/uk/discover/animals/general-animals/spotted-hyena-facts/

Learn more about these dangerously hungry mammals.

Vulture

animals.sandiegozoo.org/animals/vulture

Read more about these amazing birds and their disgusting food!

Publisher's note to educators and parents: Our editors have carefully reviewed these websites to ensure that they are suitable for students. Many websites change frequently, however, and we cannot guarantee that a site's future contents will continue to meet our high standards of quality and educational value. Be advised that students should be closely supervised whenever they access the internet.

INDEX